Fossil News

The Journal of Avocational Paleontology

Vol. 22.3— Fall 2019

4-8 **BEAUTY IN THE STONE**
A master preparator on the step-by-step process of transforming a fossil find

ROADSIDE AMERICA 10-14 **THE JOHN MARGOLIES PHOTOGRAPHIC ARCHIVE**
The eccentric highway paleoart that's going ... going ... practically gone!

NANDO MUSSMARA 17-25 **Tracking Dinosaurs at the Rioja Center for Paleontology**

WENDELL RICKETTS 26-33 **The World-Famous Dinosaur Trackways of the Connecticut River Valley**
Edward Hitchcock and the founding of the Beneski Museum of Natural History

ERIK TIHELKA 34-43 **The Origin of Insect Wings**
An interview with controversial Czech paleoentomologist Jarmila Kukalová-Peck

45 **USPS Issues** *Tyrannosaurus rex* **"Forever" Stamps**

SETH SORENSEN 47-51 **An excerpt from *Head Full of Rocks: The Mostly True Tales of a Commercial Paleontologist***
A selection of short stories from the forthcoming book

FRONT COVER: Eubrontes *sp., the ichnogenus assigned by Edward Hitchcock to Pliny Moody's so-called "Noah's Raven" dinosaur trackway, discovered in South Hadley, Massachusetts, in 1802. Because Hitchcock originally believed the tracks had been made by a prehistoric bird, he first assigned them to the ichnotaxon "Ornithichnites."* BACK COVER: *The Connecticut River Valley at the time of the dinosaurs. Mural at the Beneski Museum of Natural History in Amherst, Massachusetts. Photograph © Roger L. Bilisoly, all rights reserved; used by permission.*

Contributors To This Issue

ERIK TIHELKA (interviewer) was born in Brno, Czech Republic, and is currently a student at Hartpury College in the United Kingdom. He works on fossil deposits with exceptional preservation such as Burmese and Baltic amber to better understand the evolution of insects, with an emphasis on beetles. He is interested in the earliest terrestrial ecosystems during the Silurian-Devonian and their arthropod fauna.

JOHN MARGOLIES (photographer) was an architectural critic and expert in American "vernacular architecture." Born in New Canaan, Connecticut in 1940, Margolies became interested in roadside attractions as a child, riding in the backseat of his parents' car on trips. When he was 16, he obtained his driver's license and began sightseeing in a 1948 Oldsmobile. He began photographing in 1970 and ultimately authored numerous articles on architecture and books of his photography. Margolies was a Guggenheim fellow and received many other grants and awards for his work. He died in 2016.

JOHN SIBBICK (paleoartist) has been an independent illustrator for more than forty years, covering such subject as mythology, fantasy, and natural and prehistory. His reconstructions have been seen in many museums and traveling exhibits, including *Dinobirds* for London's Natural History Museum and commissions for the National Museum of Scotland and the Gamagori Museum in Japan, among many others. He creates drawings for production companies to aid sculptors in building both static and animatronic models of dinosaurs and other extinct reptiles. He is now at work on the fourth edition of *Dinosaurs: A Concise Natural History* by David Fastovsky and David Weishampel. He lives on the Isle of Wight, also known as "Dinosaur Isle" for the fossil bones and trackways found there.

NANDO MUSSMARA (writer) is the founder of *PaleoNature: The Multilingual Natural History Magazine* and a long-time fossil collector, science writer, and photographer. He has been a Contributing Editor to *Fossil News* since 2016. Visit PaleoNature at www.paleonature.org.

SETH SORENSEN (writer) grew up in the small town of Spring City in the middle of nowhere Utah. He is the author of *The Utah Story* and *History's Aquarium: Treasures of Fossil Lake*. The co-owner of American Quarry in Kemmerer, Wyoming, he lives in Salem, Utah, with his family. His credits his parents for his passion for rockhounding.

HOW TO SUBSCRIBE: Domestic US rates: $50.00 USD for four quarterly issues, payable via PayPal or Zelle (both via FossilNews@FourCatsPress.com) or Venmo (@FourCatsPress). Outside the U.S., please contact us for a quote. Single issues: $15.00 USD (plus postage if outside the US). Complete info at: tinyurl.com/fnsubscribe.

DISPLAY ADVERTISING: Subscribers receive 10% off display advertising. Our current "Rate Sheet & Specifications" are online at: tinyurl.com/fnadvertise.

SUBMISSION GUIDELINES: *Fossil News* welcomes submissions from all those with an interest in paleontology and fossil collecting. Submissions and queries to: FossilNews@FourCatsPress.com.

HOW TO SUBMIT YOUR WORK: Please read our guidelines (tinyurl.com/fnguidelines) or contact us for details.

BOOK REVIEWS: *Fossil News* does not accept unsolicited or previously published reviews. It's always best to query first. If possible, send clips of your published work.

EDITING AND REVISION: *Fossil News* reserves the right to edit submissions for length, grammar, punctuation, structure, and style.

PAYMENT: We offer contributors a complimentary copy of the issue in which their work appears and a choice of (a) two insertions of a display ad (up to one-half page) or (b) 50% off an annual or gift subscription.

Fossil News
The Journal of Avocational Paleontology
FossilNews@FourCatsPress.com
www.FossilNews.org

Fall 2019: 978-1-7320442-9-6

BEAUTY IN THE STONE

A master preparator on the step-by-step process of transforming a fossil find

The quarries near the small (population 2,747), high-desert town of Kemmerer, Wyoming, are world-famous for the exceptional fossils of animals, plants, and insects that lived in or near the shores of "Fossil Lake" during the Eocene—and especially for the fossil fish they yield.

Fossil Lake is one of the three lakes that made up the Green River Lake System—Lake Uinta, Lake Gosiute, and Fossil Lake. They were the "great lakes" before the Great Lakes existed.

While the lake system took thousands of years to form and lasted, in the case of Lake Uinta, for more than ten million years, Lance Grande, author of The Lost World of Fossil Lake: Snapshots from Deep Time, *notes that Fossil Lake lasted only about two million years, and the fossil-bearing deposits that bring caravans of people to Kemmerer each season—the Fossil Butte Member of the Green River Formation—were probably formed over no more than a few tens of thousands of years.*

In that tiny geological sliver of time, the organic remains of the turtles, crocodiles, plants, lizards, horses, insects, bats—and millions of fish—that have been found in the Kemmerer quarries began the long, slow process of becoming the fossils that are so highly prized today.

But not everything that comes out of the ground—or from between layers of rock—is in perfect shape. And that's where preparators come in. People in and around Kemmerer have been prepping—that is, cleaning and restoring—Fossil Lake fossils for well over a century, starting with men like David C. Haddenham and Robert Lee Craig, two of America's early full-time fossil quarriers. According to local lore, during the decades when Craig worked the quarry (1897-1937), he prepped fish by lantern light with a penknife and sold them at the train station in the town of Fossil, about eleven miles west of Kemmerer. Fossil is "the town that was" (all that remains are the crumbling ruins of the station itself and a few assorted shacks—though efforts are underway to restore the area for tourists), but the Craig quarry is still producing and is now owned and operated by the Green River Stone Company.

Anyone who has seen beautiful fossils in museums, galleries, or friends' collections has seen the result of preparators' labor, but how fossils get from "that" to "this" usually takes place behind the scenes. Since Haddenham's and Craig's day, many preparation techniques have become extremely sophisticated, but there's still no substitute for the putting in the long, patient hours needed to bring a fossil slowly and painstakingly to light or to restore it to the potential that sometimes only a master preparator can see hidden in stone.

Robert Bowen of Wyoming Fossils is one of Kemmerer's most accomplished preparators, and fossils he's prepared are now in collections all over the world. When a client brought him a specimen of a freshwater stingray, Heliobatis radians, *collected at the American Fossil Quarry near Kemmerer—in multiple pieces—Bowen set to work. He shared with* Fossil News *the step-by-step process of preparing and restoring the fossil into something that anyone would be proud to display at home. The specimen had been discovered in the so-called "split fish" layer, meaning Fossil Lake's relatively softer sediments in which slabs of rock commonly split cleanly through a fossil, leaving a part and a counterpart on separate halves of the slab. That's what had happened with the stingray found by Bowen's client.*

From top left: *1. The specimen as it came out of the quarry. Both the part and counterpart are in pieces, and chips have broken loose from the surfaces in several places. 2. The restored stingray will eventually be inlaid into a new stone slab. In this case, the slab that will hold the specimen also contains a small fish. The first step is to trace the outline of the slab on an oak plywood backing, which will strengthen and support the fossil. 3. Bowen cuts carefully around the outline with a jigsaw. 4. One side of the slab is treated with Liquid Nails,® a heavy-duty adhesive used in the construction industry. Once dry, Liquid Nails® can be sanded and painted. 5. The adhesive has been smoothed to cover the entire surface, and the host slab can now be carefully pressed onto the plywood. The display surface of the slab is protected, weight is placed on top, and the adhesive is left to cure for at least twenty-four hours.*

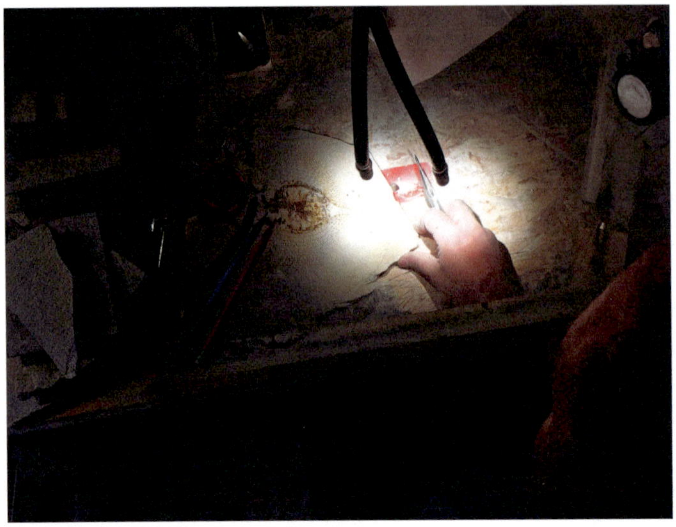

From top left: *6. Before proceeding with assembly, Bowen cleans the individual pieces with an airjack and small hand tools under a Plexiglas hood. 7-8. Now comes the first real attempt at fitting the jigsaw puzzle together. What is quickly obvious is that, in reality, the specimen is approximately two thirds of a stingray, with most of one wing missing. A portion of the specimen is on the part and another portion is on the counterpart (sometimes called the "positive" and "negative"). When the part and counterpart are laid side by side, it begins to be possible to visualize how an entire stingray might be assembled from these pieces by "cloning" the counterpart. 9. The pieces are trimmed, thinned to a uniform thickness, then fitted together and glued. 10. The disc of the stingray body begins to take shape.*

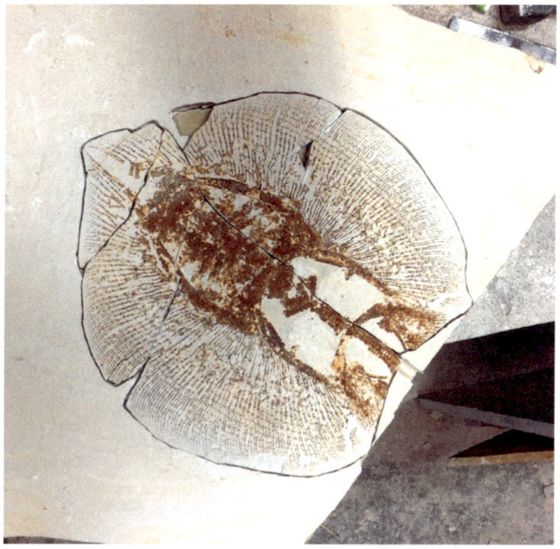

FClockwise from top left: *11. Closeup of the ray's placoid scales (or dermal denticles), tiny, teeth-like structures that streamline rays and sharks for swimming. 12. The disk has been fully reconstructed and is placed on the host slab to be inlaid. An outline is drawn in pencil around the fossil to indicate the area that will be routered out to receive it. Because the specimen is missing most of the tail, Bowen has chosen to position the ray so that it appears to be swimming onto the slab. 13. The area where the stingray will be inlaid has been routered to the desired depth. The screws/wire are used to lift or move the specimen without risking damage to the fossil. 14. The ray fits nearly perfectly into the space created for it. Glue is next injected into all openings to secure the specimen permanently to the host rock. 15. When the glue is dry, PaleoSCULP® resin by PaleoBond is painstakingly worked into all gaps to seal them completely.*

From top left: *16-17: Once all the gaps have been filled, the specimen is ready for painting and touch-up: the restoration part of the process. The resin infill is colored to match the host rock, and any missing or damaged parts of the fossil can be enhanced with slow and careful work using a thin artist's brush. Some clients prefer to skip this step. 18. All but done. A few final tweaks remain to blend colors and make the specimen look its best, and then the exposed back surface of the oak base will get a coat of varnish. At that point, the repaired and restored specimen will be ready to hang on the wall or display on a stand.*

WYOMING FOSSILS

Sawaya's Center
921 Pine Avenue
Kemmerer, Wyoming
(254)223-3204

visit our website www.wyomingfossils.com

OPEN DAILY
9:00 a.m.
to 6:00 p.m.

- Fossils
- Minerals
- Crystals
- Unique Items
- Gifts

❖FIND US ON FACEBOOK/INSTAGRAM UNDER WYOMING FOSSILS

ROADSIDE AMERICA
THE JOHN MARGOLIES PHOTOGRAPH ARCHIVE

The eccentric highway paleoart that's going ... going ... practically gone!

When John Margolies died in May 2016 at the age of 76, he left behind the dozen or more books he had written; the vast collection of postcards, pennants, matchbooks, travel brochures, diner place mats, maps, do-not-disturb signs, and menus he had amassed in forty years of travel across the forty-eight contiguous states; and nearly 12,000 color slides of scenes from "roadside America" that he lovingly and painstakingly photographed, beginning in the mid-1970s, with nothing more than a standard-issue, 35 mm Canon FT camera.

In all those years, Margolies, a native of Connecticut, covered a lot of ground (more than 100,000 miles is a conservative estimate), but the distance he traveled was a great deal more than geographical. When he started out, he was an amateur photographer on a quirky personal mission to photograph what he considered an endangered species—the mini-golf courses, roadside restaurants, drive-in theaters, barbershops, billboards, gas stations, decaying movie palaces, and faded RV parks that were once a common sight along America's highways and were already disappearing when Margolies started documenting them. By the later decades of his life, however, he was widely praised as an architectural historian and an expert in what is known as American "vernacular architecture," his services as a lecturer were in high demand, and his photographs were exhibited in galleries and museums around the world.

The titles of Margolies's illustrated books reveal a great deal about the subjects that interested him most: *Pump and Circumstance: Glory Days of the Gas Station* (1993); *Home Away From Home: Motels in America* (1995); and *Ticket to Paradise: American Movie Theaters and How We Had Fun* (1995)—just to name three. To capture images exactly as he wanted them, Margolies stayed in sketchy motels and rose before dawn to take advantage of the early-morning light—and the absence of other human beings. He waited patiently for clouds to pass (the blue of the background skies in his images is often pristine), and his long-time partner, Jane Tai, remembered that he "carried a broom in his car, and ... would do a little clean-up so that there wasn't any extra debris in the frame."

By 2009, the value of Margolies's work, and the history it rexorded, was recognized by the Library of Congress, which began buying Margolies' photography collection; the bulk of it arrived in 2015. (The Library also acquired Margolies's vast accumulation of "ephemera"—more than 3,800 items relating to American travel and tourism, including postcards, pamphlets, brochures, catalogs, travel logs, advertisements, maps, and sheet music).

Digitized, organized, and tagged for searching by the Library's Photography, Prints, and Photographs Division, Margolies's work has become the John Margolies Roadside America Photograph Archive, a immense collection now available to the public online. The Margolies archive is:

> one of the most comprehensive documentary studies of vernacular commercial structures along main streets, byways, and highways throughout the United States in the twentieth century.... Margolies' Roadside America work chronicled a period of American history defined by the automobile and the ease of travel it allowed. Emerging with the prosperity of the post-WWII era, roadside and commercial structures spread with the boom of suburbanization and the expansion of paved roads across the United States. Yet, in many instances, the only remaining record of these buildings is on Margolies' film,

Facing page: *A somewhat fanciful creature, apparently part glyptodont and part stegosaur;* and this page, top: *Thunderbeast Park, Route 97, Chiloquin, Oregon, 1987.* This page, middle: *The "Mammoth Hole at Holiday Golf, Daytona Shores, Florida, 1985.* Bottom*: Petrified Creature Museum, Richfield Springs, New York, 1995. Credit: John Margolies Roadside America Photograph Archive (1972-2008), Library of Congress, Prints and Photographs Division.*

Bob's MUSEUM
ROCK and GIFT SHOP
WHOLESALE JEWELRY
GIFTS
PINS
MARBLES
ONYX EGGS 1.50
MOTOR OIL 1.25
TRAINS $5.00

ROCK SHOP

Preceding pages, verso, top: *Bob's Museum, Rock and Gift Shop, I-84, Bliss, Idaho, 2004.* Bottom: *Indian Rock Shop, Holbrook, Arizona, 2003.* Preceding pages, recto: *the "Green Dinosaur Hole," Wacky Golf, Myrtle Beach, South Carolina, 1979.* This page, at left, top: *Petrified Creature Museum, Richfield Springs, New York, 1995.* Center: *Dinosaur Village RV Mobile Home Park, Route 40, Jensen, Utah, 1991.* Bottom: *Dinosaur Garage, Dinosaur, Colorado, 1991.*

because tourist architecture was endangered by the expansion of the interstate system and changing travel desires.

Today, much of that architecture is worse than endangered today— it's gone for good. In fact, it was already on its way out when Margolies immortalized this quintessentially American style on film: He sometimes arrived just days before a building was demolished.

Among the charming curiosities in the prodigious Margolies collection are his images of roadside representations of prehistoric beasts: brontosaurs and other dinosaurs, saber-toothed cats, ancient giant mammals, and, in at least one case, a fanciful creature that seems be to a cross between a glyptodont and a stegosaur. These are a few of the photos of the roadside American paleoart that Margolies captured in his travels.

Note

[1]Fox, Margalit Fox (2016, 2 June). John Margolies, Photographer of Whimsical Architecture, Dies at 76. *New York Times*. https://www.nytimes.com/2016/06/03/arts/john-margolies-photographer-of-whimsical-architecture-dies-at-76.html.

Nando MUSMARRA

Tracking Dinosaurs at the Rioja Center for Paleontology

The date is 114.9 million years ago, give or take a few millennia, and the morning air is still and damp.

A baby iguanodon, under the watchful eyes and congenial snouts of others of its group, strolls along the banks of a river in search of tasty fern fronds. Nothing seems to disturb the valley's calm. Alongside the group of iguanodons, a vast sauropod casts a refreshing shadow, The baby ornithopod entertains itself by chasing the moving patch of shade over the muddy ground, in part for the fun of it, and in part for relief from the heat and humidity of a new day in the Lower Cretaceous.

A gust of wind from a nearby grove of conifers carries a pungent, gamy odor toward the river, flustering the group of iguanodons. Their hearts begin to pound, and their fear grows. For a moment, they freeze and then, sensing that mortal danger is suddenly near, they scatter, the sound of their thundering panic breaking the silence. Overhead, a flying reptile shrieks irritably, adding to the chaos. The giant sauropod, despite its bulk, is the first to flee.

From out of the conifers emerges the elongated head of a Baryonyx, *which gazes menacingly toward the herd. Another appears from the opposite direction, coming up behind the iguanodons in a surprise attack. That perfectly executed pincer maneuver could easily have sent the adults running for their lives, isolating the baby iguanodon and handing the spinosaurs an easy meal. Instead, they turn their backs and walk closer to the water, attracted by the sight of a fat* Lepidotes *swimming near the surface, easy prey they can scoop from the river with their hooked claws. When all is finally quiet again, their tracks remain in the soft mud of the river bank.*

One hundred and fifteen million years later, that fossilized mud still preserves signs of the passing of *Baryonyx* and other dinosaurs as trace fossils—tracks and trackways—that can be seen in exposures near Igea, a small town in Rioja Province, in Spain.

Igea has gained a certain notoriety for its paleontological riches (several thousand dinosaur footprints have been found around the town), and its residents are quite proud to walk the same ground that dinosaurs once trod. A school teacher and his fellow citizens took it upon themselves to gather local minerals and fossils for display and study and, over the years, their excellent collection has grown and expanded. In time, it found a permanent home: the Municipal Museum, which is also the site of the Rioja Center for Paleontology (Centro de Interpretación Paleontológica de La Rioja).

A Environment of Lakes & Frequent Floods

Displays at the entrance to the Rioja Center include geological maps and large dioramas that help explain the paleogeographic context of Rioja Province during the Upper Jurassic and the Lower Cretaceous. At the time, the Iberian plate was located between the twentieth and thirtieth parallels in a sub-tropical zone with a warm, humid climate. Over and over during millennia, periods of plate movement alternated with frequent major floods that contributed to the rapid accumulation of sediment.

The exposure of dinosaur tracks in Rioja Province belongs to the Leza and Jubera Formations. The first is largely composed of limestones deposited in ancient lakes that once existed in the valley of the Leza, San Vicente di Robres, Arnedillo, and Préjano Rivers. Study of fossils of freshwater green algae have dated those sediments to approximately 114 million years BCE (Lower Aptian). The limestone layers alternate with sandstone, clay, and marls, and pyrite is present, a sign that the ancient bodies of water were stagnant and anaerobic. The Jubera Formation, made up of conglomerates, sand, and reddish clays, conversely, has its origins in ancient flood basins.

Rocks of these types suggest that the Riojan dinosaurs once walked on a wide river plain that was crisscrossed

Facing page: "*Overhead, a flying reptile shrieks irritably, adding to the chaos....*" © *Loana Riboli; used by permission.* Above: *A life reconstruction of* Baryonyx *and potential prey in one of the dioramas at the Rioja Center for Paleontology. This and all photos in this article, unless otherwise noted © Nando Musmarra; used by permission.* Following page: "Baryonyx" *by John Sibbick. © John Sibbick; all rights reserved. Used by kind permission of the artist.*

SIBBICK 95

1 YEAR ANNIVERSARY SALE
For Online Store
PaleoJoe: The Fossil Store
Museum and High Quality Fossils
One of a kind fossils
Dinosaur Bones, Dinosaur Teeth, Trilobites, Fish, Crinoids, Ammonites, Ferns, Cystoids, Dinosaur Posters, Shark Teeth, Childrens books ....so much more.
"Thank you for the speedy delivery. They look great along with the other 18 specimens I have." -David, TX
"Thanks. I love my new Trilobite, and that beautiful matrix it is in, helps make it the perfect piece to liven up my boring office cubicle" -Lawrence, Michigan
"I have just got the fossil. It is superb!" -Manuel, Brazil
"I really love your products and business practices" -Janice, West Virginia
Shop PaleoJoe.com The Fossil Store
Enter code: 1year for a 15% OFF Sale to celebrate our 1st year of business.
All orders over $100 include free shipping in the USA
Contact Jen at jenkchodl@yahoo.com or PaleoJoe at paleojoe@charter.net or visit the website PaleoJoe.com for more Fossil Fun

by a multitude of lakes and channels. As a result of frequent and significant flooding, many areas were often covered in water, the ideal substrate to preserve large numbers of tracks in damp or muddy soil.

The Rio Cameros Valley, located between the river basins of the Ebro and the Douro Rivers, looked very different during the Lower Cretaceous than it does now. The creatures of the air and skies were not griffons and vultures, but rather pterodactyls and, on terra firm, dinosaurs grazed and hunted instead of today's wild boars, cows, and goats. In the area's waters, where lakes (some of which were connected to the Tethys Ocean), rivers, and seashores shared a wide coastal delta, *Hybodus* sharks and other large fish such as *Lepidotes* swam. In the swamps, connected by a myriad of small canals, crocodiles and other reptiles wallowed. The meeting of ocean, freshwater, and brackish environments allowed a remarkable number and diversity of species to proliferate. They included freshwater and marine mollusks (*Unio* and *Eomiodon*), the gastropods *Glauconia* and *Wealdenia*, fish, and reptiles and amphibians, to name a few. (The Center's collection of complete specimens of *Lepidotes*; the teeth and fin spines of the extinct shark, *Hybodus*; and fossils of turtles, tortoises, and crocodiles are especially beautiful.)

A large diorama on the second floor reproduces the Monte Santa Cruz exposure, where the fossilized trunks of the extinct tree fern *Tempskya riojana* have been found. These fossils, club- or cone-shaped as is typical of *Tempskya* (the "trunk" was actually a bundle of stems, which has led them to be called "false trunks"), are rounded at the ends, a testament to the significant erosion the plant remains underwent as they were transported over long distances by river currents or in floods. They are some of the few, though important, traces of the vegetation that once grew in Rioja Province.

Paleoichnology & the Study of Dino Trackways

The Center's most spectacular rooms, though, are dedicated to paleoichnology—trace fossils. Specimens of natural casts of theropod footprints abound, and exhibits show reconstructions of two nearby localities where dinosaur tracks have been found: the La Torre site, accessible only to skilled rock-climbers, where an amazing, well defined sequence of tracks made by a running theropod is exposed; and the Era del Peladillo site, from which the very first ichnofossils of webbed footprints (ichnogenus: *Hadrosaurichnoides igeensis*) were described in 1992 by paleontologist María Lourdes Casanovas-Cladellas. Though Casanovas-Cladellas interpreted the tracks as having been made

Facing page, top: Lepidotes elvensis *skull. Photo by Jeff Kubina, courtesy of WikiCommons.* Bottom: *Natural cast of a large ornithopod track, possibly an iguanodontid. © Nando Musmarra; used by permission.*

10 cm
1 2 3 4 5 6 7 8 9

Facing page, top and middle: *Dinosaur "ichnites" (fossil footprints) from the Peñaportillo Formation, Munilla (La Rioja, Spain. España).* © *Ignacio Benvenuty Cabral and the Banco de Imágenes Geológicas. Bottom: Tracks of* Baryonyx *at La Rioja.* This page, top: Baryonyx *reconstruction at the Centro de Interpretación Paleontológica de La Rioja, Igea.* Bottom*: The formation of a dinosaur footprint, from the La Rioja Museum.*© *Nando Musmarra; used by permission.*

A Spanish Baryonyx

For the study of the *Baryonyx* in the Center's collection, Dr. Angela Milner of London's British Museum, the first scientist in the world to publish a study of *Baryonyx*, came to Igea. She confirmed the identification made by Spanish paleontologists Angela Torres and Luis Ignacio Vieira: the Igea specimen is a *Baryonyx*, a carnivorous dinosaur with an elongated skull somewhat like a crocodile's. *Baryonyx* sported some thirty-two pointed teeth per jaw, and the shape of its skull helped the animal search for food under water.

Baryonyx, the super-predator at the top of the food chain of the Rioja delta, was an impressively large dinosaur, and apparently grew to from six to ten meters in length (more than nineteen to more than thirty-two feet) and was between three and four meters (roughly ten and thirteen feet) tall. It may have weighed some 2,000 kilograms

by an ornithopod (a member of the "bird-hipped" Order Ornithischia), work a decade later redefined them as theropod trackways.

Carnivores Had the Upper "Hand"

A calculation of the number of dinosaur tracks found in the Igea area reveals that carnivorous dinosaurs far outnumbered herbivorous species (in a ratio of eighty to twenty). According to paleontologists, this unusual prevalence of carnivorous dinosaurs can be explained by the exceptional abundance of fish, which were probably the key to the proliferation of carnivorous species. Conversely, the scarcity of plant fossils suggest that there was insufficient vegetation to satisfy the needs of herbivores, limiting their populations (the few examples of plant fossils found in the area so far include trunks transported over long distances by flooded rivers and one complete trunk of *Dadoxylon (Araucarionxylon) riojense*, a conifer, found at the La Eremita de Nuestra Senora del Vilar site.)

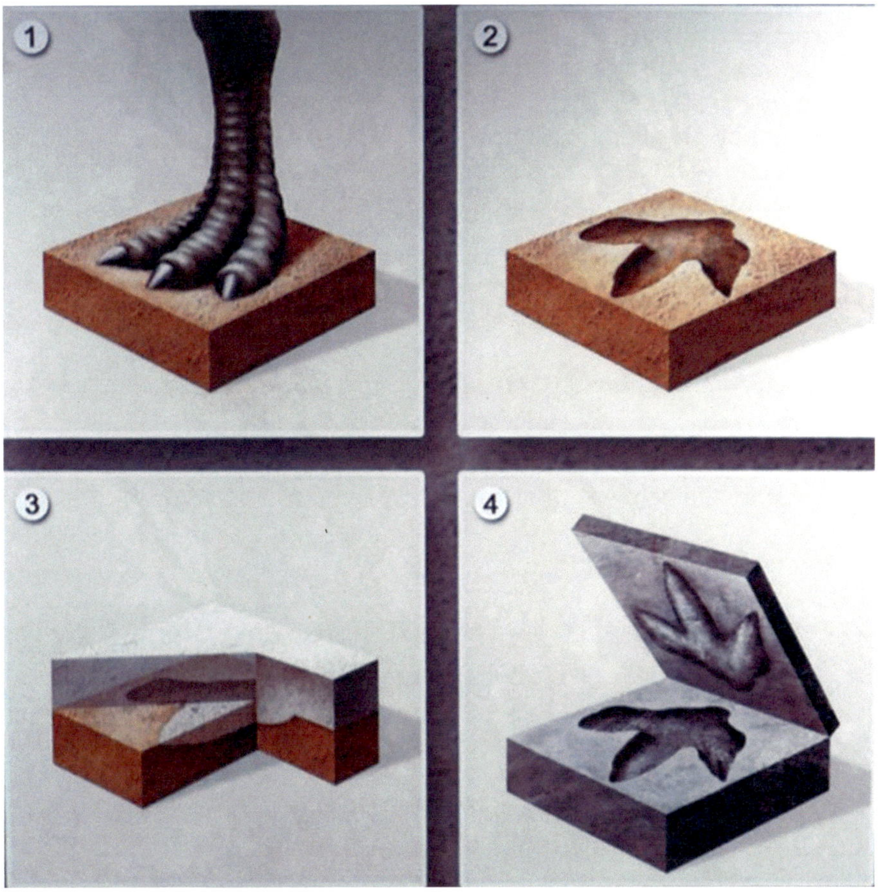

(more than two tons). The Igea specimen shows the sharp claws that allowed *Baryonyx* to skewer large fish, exactly like the specimen discovered near Surrey, England, in 1983, in whose stomach the giant scales of *Lepidotes* had also been fossilized.

Baryonyx is known largely for the chance finds of incomplete remains. The Spanish *Baryonyx*, found by amateur paleontologist Pachi Sanchez, is the second in the world to be subjected to official study and publication in 1994. More complete than the English *Baryonyx*, Sanchez's specimen preserves a foot as well, which shows evidence of a paleopathology: a fractured fibula that later healed while the animal was still alive. The dinosaur likely survived this serious injury because of the great abundance of fish in its environment, a kind of "home delivery" of the food it needed to appease its hunger.

— *Nando Musmarra; English translation by Wendell Ricketts*

The Science of Dinosaur Tracks

The study of dinosaur tracks considers all aspects of the track or print, including dimension and shape, distance between one footprint and another, and the depth of the track. Often, identifying the species of dinosaur that left a particular track is difficult, but it may be possible to determine at least the family or other group to which it belonged. Study of tracks and trackways allows scientists to infer a great deal about the living dinosaur, including weight, size, limb length, and even speed. These measurements permit the animal's biomass to be calculated—and even make it possible to estimate its daily calorie and energy requirements. Dinosaur-track ichnology also helps paleontologists formulate hypotheses about the social habits of dinosaurs—for example, whether they hunted alone or in packs, whether they migrated in flocks, or whether they cared for their offspring.

Sauropod Tracks

Spanish paleontologists distinguish the tracks of sauropods from one another not chiefly on the basis of their size, but by the varying positions of the limbs in a "manus-pes" (forelimb-hindlimb) sequence. They're divided into Type S1 (*Brontopodus*) and Type S2 (*Parabrontopodus*)tracks.

Brontopodus-type tracks indicate were left by very large quadruped dinosaurs, and the diameter of hindlimb tracks can be greater than 60cm (just under two feet). The details of digits (fingers and toes) are unlikely to be preserved. The crescent-shaped imprints of their forelimbs are smaller. *Brontopodus* tracks were made by an animal with a wide gait with a lot of distance between right and left tracks.

Parabrontopodus or Type S2 imprints are morphologically similar to the S1 type, but the gait of the animal that left them was "narrower" and left little space between right and left tracks.

Theropod Tracks

The tracks of theropod dinosaurs are the most common in the Rioja deposits. They're divided into three types.

T1 tracks are large, three-toed imprints of a creature that walked bipedally in a "digitigrade" style (meaning on its toes). The digits are robust and pointed, and the third digit is sometimes turned inward. These tracks sometimes preserve the impression of the heel. Tracks of the T1 *Buckeburgichnus maximus* type were presumably made by large carnivorous dinosaurs like *Baryonyx*.

T2-type tracks are also three-toed and resemble the previous type, but they are clearly smaller (15-30cm.). The impressions were left by relatively robust digits morphologically similar to the T1 type, though the shape of the heel is more slender. The animals that left these tracks were bipedal with a quite narrow gait. The likeliest candidates for T2 tracks are medium-sized theropods like *Velociraptor* or *Deinonychus*.

Type T3 tracks are small and three-toed, with thin, elongated, and occasionally curved digits. Scientists believe they were left either by small theropods or by the juveniles of larger species.

— Nando Musmarra; English translation by Wendell Ricketts

The World-Famous Dinosaur Trackways of the Connecticut River Valley

Edward Hitchcock & the Founding of the Beneski Museum of Natural History

From the outside, the Beneski Museum of Natural History, located on the campus of Amherst College in Amherst, Massachusetts, is a staid and unprepossessing three-story, red brick building, rather less stately than many of the other buildings on the college's nearly two-hundred-year-old campus.

Inside, however, the Beneski is home to an exciting and impressive collection of fossils and minerals collected locally and abroad, including one of the world's largest and most studied collections of fossil dinosaur tracks, the Hitchcock Ichnology Collection.

The Ichnology Collection was originally known as the "Appleton Cabinet," after the name of the donor who contributed the funds for its construction in 1855 (at the time, natural history collections were frequently known as "cabinets of curiosities"), and later renamed for Edward Hitchcock, a geologist and Congregationalist minister who served as professor of Natural Theology and Geology at Amherst College from 1825 until 1864 and, from 1845 until his death, as the College's third president. The Appleton Cabinet was constructed to house Hitchcock's personal ichnology collection as well as the numerous zoological and other scientific specimens that Hitchcock encouraged alumni to send back from all over the world.

Today the Museum houses roughly 200,000 objects, including more than 1,700 slabs containing some 21,000 fossilized tracks and traces, one of the largest collections in the world, the majority of which Hitchcock collected himself, beginning in 1835, from sandstone deposits across the Connecticut Valley. The Hitchcock Ichnology Collection is justly famous for its dinosaur footprints, but it also contains a large number of tracks and other traces of small reptiles and amphibians as well as invertebrate trails.

Noah's Raven

Perhaps the most historically important specimen in the collection is the so-called "Noah's Raven," a trackway that ultimately turned out to be the first North American evidence of the presence of dinosaurs. In fact, the

Facing page: *One of the 1,700 slabs containing tracks of dinosaurs, other reptiles, amphibians, and invertebrates at the Beneski Museum. Image © Roger L. Bilisoly, all rights reserved; used by permission.* This page, left: *Engraving of the "Moody Foot Mark Quarry, South Hadley, Massachusetts." Illustration from Edward Hitchcock's* Ichnology of New England, *1858.* Right: *The famous "Noah's Raven" slab, found in 1802 by Pliny Moody, on display at the Beneski. Image © Roger L. Bilisoly, all rights reserved; used by permission. (Unless otherwise mentioned, all photos from the Beneski Museum are © Mr. Bilisoly.)*

trackway had been found in South Hadley, roughly ten miles south of Amherst college, in 1802. Its discoverer was a twelve-year-old farm boy named Pliny Moody who, according to local legend, had been plowing a field when the slab was exposed. He dragged the 19" x 31" sandstone slab home, and his family installed it as their front doorstep. There it remained for years until it was bought by a local doctor, eventually coming into Hitchcock's possession around 1839. Hitchcock first assigned the Moody tracks to the ichnotaxon "Ornithichnites" and later added it to the Ichnology Collection with accession number "16/2."

The "Noah's Raven" slab is so named because (or so the story goes) Moody or his contemporaries would tell people that the marks were the tracks of the raven released by Noah from the Ark at the end of the biblical Flood. The raven, they said, had landed in what would later become Western Massachusetts. The specimen is technically the first North American evidence of dinosaurs. For nearly forty years, however, no one outside of the Moody family knew of its existence. In the meantime, another trackway was found, and so credit for the first find officially known to science goes to Dexter Marsh, a laborer who, in 1835, was hired to lay sidewalk slabs in Greenfield, Massachusetts, with flagstones from a nearby quarry.

Marsh noticed tracks on one of the slabs, and word quickly spread in the neighborhood about the discovery of "petrified bird tracks." Marsh's specimen soon came to the attention of a local doctor and avocational scientist, James Deane, who immediately wrote to both Hitchcock at Amherst and to Benjamin Silliman at Yale, sending sketches of the impressions and asking their opinions. Neither man believed Marsh had found fossilized bird tracks. After several more exchanges of letters, however, Deane eventually convinced Hitchcock—who had, in the meantime, also been encouraged by Silliman—to come see for himself. Hitchcock was instantly converted: they were bird tracks, he said, and he quickly began preparing his first scientific article on the study of what he called

"ornithichnology," meaning the "study of fossil bird traces." Marsh's specimens were added to the museum's collection with acquisition numbers 18/1 and 18/2 and, in time, they were given the name *Tridentipes elegans*.

That Hitchcock originally misidentified the fossils as the tracks of birds isn't surprising. At the time he began his studies, the existence of dinosaurs—and even the word "dinosaur"—was unknown. For all his curiosity about science, however, Hitchcock never stopped believing in the Biblical story of creation and was critical of his contemporaries, Charles Darwin and the prolific Swiss-American biologist and geologist, Louis Agassiz, another creationist.

A Less Than Enthusiastic Reception

Hitchcock worked quickly to get an official report of the Connecticut River Valley "ornithichnites" into print, and published his first article about them in the *American Journal of Science and Arts* only a year after the discovery of Marsh's trackway. Though several important scientists supported his analysis of the tracks, a larger number of important geologists and other scientists of the time were doubtful, including most of the membership of the Geological Society of London. Hitchcock's evidence was unconvincing, they said, adding that nothing could be said conclusively until body fossils were found of the "giant birds" (one track was seventeen inches long) who had left their prints in mud of the humid swamps and lowlands that once covered western Massachusetts.

Hitchcock, however, was undaunted, and continued to defend his ideas and to publish additional articles and monographs about the "ichnology of New England," the

This page: *A large three-toed track on display at the Beneski.* Facing page, top right: *Another trackway, colored for visibility, shows the print of a large, three-toed individual and the smaller tracks of a different animal. Both © Roger L. Bilisoly, all rights reserved; used by permission.* Bottom: *An engraving of the fossilized tracks of what Hitchcock believed was a giant bird. The illustration was drawn by Orra White Hitchcock, one of America's earliest women botanical and scientific illustrators and artists. It appeared in Edward Hitchcock's 1836 description of strange, three-toed prints found in the red sandstone of the Connecticut River Valley.*

title of his 1858 book. By that time, knowledge about dinosaurs was much more widespread (the word "dinosaur" had been coined in 1841 by Sir Richard Owen, and fossil footprints had turned up in Europe and Britain as well), and Hitchcock accepted the dinosaur origins of his "Ornithichnites," renaming the Moody prints *Eubrontes tuberosus*. The Connecticut River Valley tracks, which made up the bulk of the Hitchcock Ichnology Collection, were eventually all reinterpreted and identified as the traces of Triassic-age dinosaurs, of other reptiles, or of amphibians.

Marsh the Collector

Marsh remained a prolific amateur fossil collector for the rest of his life. On May 20, 1848, he wrote to Benjamin Silliman, the senior editor of the *American Journal of Science and Arts*, where Hitchcock's first article about the "bird tracks" had appeared, to describe his collection:

> You will recall that the first specimen of fossil footprints of birds ever brought into public notice in this country, was the slab I discovered among the flagging stone, while laying the sidewalk near my house.... Since that time I have felt an increasing interest in the subject, and have spent much time each year, in searching for these interesting fossils, and you will be able to judge of my success, when I tell you that I have in my collection more than eight hundred footprints of birds and quadrupeds, besides having furnished specimens to many individuals and institutions in this and other countries. I have some very perfect tracks of a quadruped so small that a five cent piece will more than cover the entire impression of the foot, and the tracks of a bird that measures more than half a yard from the heel to the point of the longest toe.... The most perfect specimens I have been able to

obtain, are from Turner's Falls, or its immediate neighborhood; they not only show the joints of the toes, but in some specimens perfectly exhibit the impression of the skin.... But some of the largest (and most perfect for large ones) I have ever seen, I obtained on the eastern declivity of Mount Tom, near South Hadley Falls. If the height of these birds was in proportion to the length of their feet, when compared to some existing birds, they must have stood twenty feet high.... I have one slab containing two footprints of a large bird ... but the great weight of the bird (probably a thousand pounds or more) pressed the sand so hard that it is perfectly smooth, showing distinctly the structure of the bottom of the foot.

Marsh's large collection was auctioned off by his family after his death in 1853.

The Dinosaur Footprints Natural Reserve

A small notice in the Electric Railway Journal of February 18, 1922, gave the news that the Massachusetts Legislature had voted to authorize the purchase of a five-acre tract from the Holyoke Street Railway, a property that included "the celebrated Dinosaur Ledge, with its tracks of the mammals of prehistoric times." The site, which included the original location of the Marsh trackways,

At left: *The sketch that Dexter Marsh included with his May 20, 1848 letter to Benjamin Silliman: "I opened a new quarry on the bank of the Connecticut River, near the mouth of Fall River ... and succeeded in obtaining two or three hundred foot prints of various birds and quadrupeds.... I only forward you a sketch of one of the quadrupeds.... You will see that this is a walking, and not a leaping animal, the fore feet are very small in proportion to the hind ones...."* Below: *Specimens in the Beneski's collection.* Facing page: *Carlton Nash, the founder of Dinosaurland, outside the business in the 1950s. © Kornell Nash. In 2017, Carlton's son, Kornell, told* Yankee *that he'd thought about giving up the business: "I could do it another 10 years, probably. At the same time, I feel like I'm in the spot where if the right person comes along, I'd pass it on." From "Twilight in Dinosaurland" by Justin Shatwell. Kornell passed in October 2019.*

languished for some time after it was purchased, uncurated, overgrown, and largely without signage or amenities. In 1936, a brief item in *Time* magazine reported that "vandals had invaded the park, cut loose more than a ton of track-bearing sandstone with pneumatic drills, [and] carted the precious material away." In fact, vandalism and theft of tracks from the reserve, and from the area generally, have continued to the present. Today the reserve is still open to the public and is known as the Dinosaur Footprints Nature Reserve in Holyoke, Massachusetts.

Another dinosaur-track site, this one near South Hadley and about a mile from Pliny Moody's 1802 find, was identified by a young Amherst College graduate, Carlton Nash, in 1933. Out of work like so many during those Depression years, Nash filled his time with odd jobs. Unlike many others, however, he had something else to fill his idle time: fossil collecting. The outcrop of trackways that he discovered remained his secret for six years until, in 1939, he was able to buy the private property on which the outcrop stood. Soon after, he opened Nash's Dinosaurland to the public, which continues today as the Nash Dinosaur Track Site and Rock Shop.

Back to the Beneski

The Beneski may not be widely known outside of the area (or by non-ichnologists), but there's a lot more to this museum than dinosaur tracks. Amherst College was a leading contributor to "dino fever" during the 1920s and houses many specimens from expeditions its scientists staffed or led. The museum also boasts an impressive prehistoric mammal collection, including full skeletons of *Brontops*, a distant relative of rhinos and horses; the bear dog *Daphoenus*; the small rhinoceros *Menoceras*; *Oxydactylus*, an extinct camel relative; and *Merycoidodon*, a pig-sized herbivore from the Eocene and Miocene, all of them collected during various Amherst College expeditions to Wyoming, South Dakota, and Nebraska during the early 1900s. Mammoth, mastodons, smilodonts, a large collection of invertebrate fossils, as well as minerals and meteorites round out the collection. Admission is free.

More Treasures of the Beneski

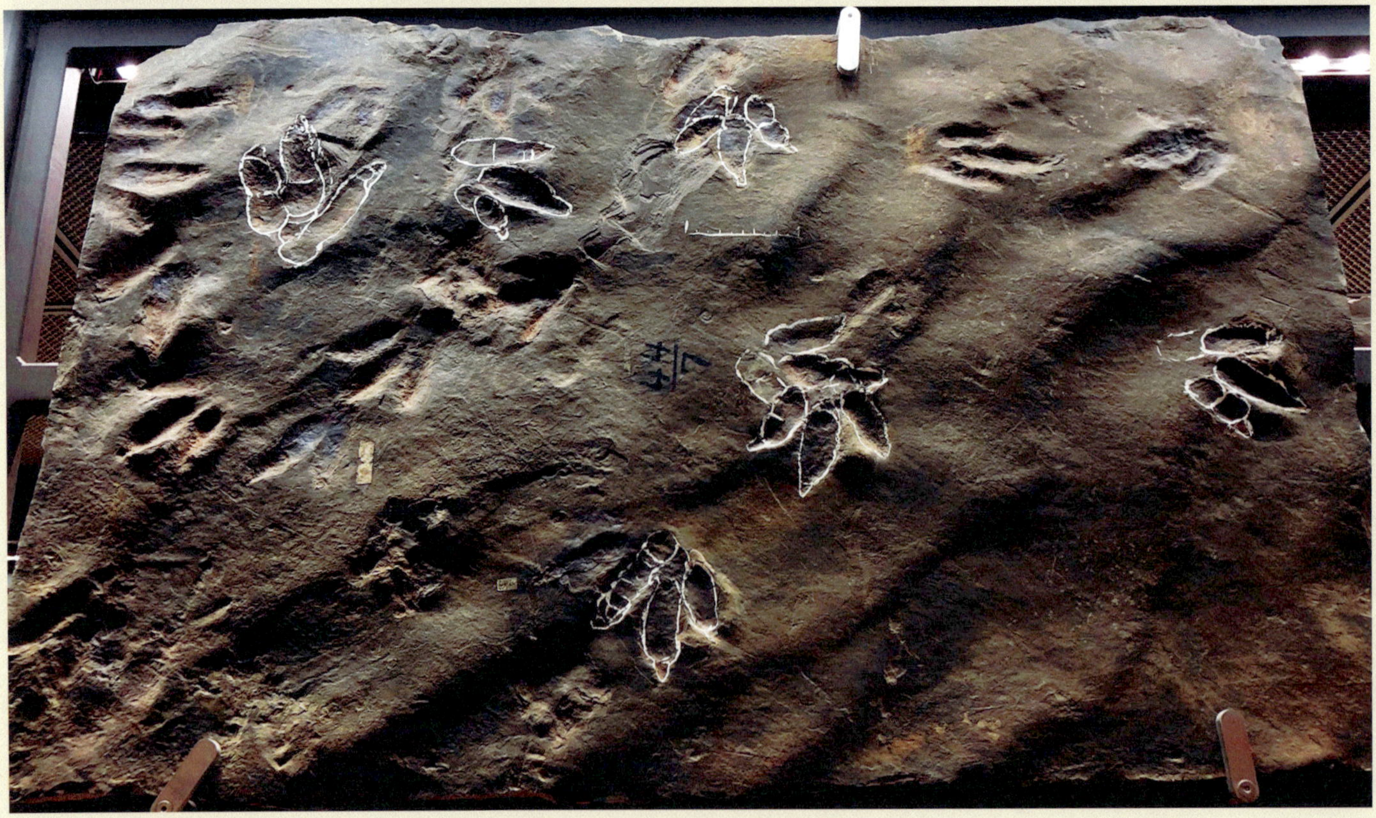

the origin of INSECT 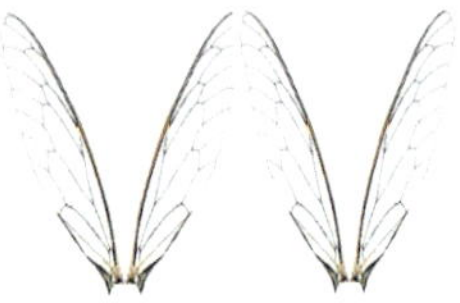INGS

An interview with controversial Czech paleoentomologist Jarmila Kukalová-Peck

by Erik Tihelka

Insects are the most diverse group of macroscopic organisms, accounting for nearly 80% of all described animal species. It is estimated that Earth is inhabited by around 4.5 million species of insects. Their stunning diversity and dominant position in virtually all terrestrial ecosystems is not a recent phenomenon. What is considered the oldest insect fossil is a springtail, Rhyniognatha hirsti, *from the Devonian Rhynie chert, and scientists discovered what they believed to be the first primitive flying insect in 2008—the full-body impression of a 300-million-year-old specimen from the Carboniferous period. Untangling the paleobiological basis of the remarkable success of insects has been the life work of Dr. Jarmila Kukalová-Peck, a Czech-Canadian paleoentomologist and research professor at Carleton University in Ottawa, Canada. A foremost specialist on the origin and evolution of insect wings and flight, Kukalová-Peck is well-known for the controversial ideas she has advanced over her long career—particularly for her widely disputed theory of the origin of insect wings. Kukalová-Peck was elected a Fellow of the Royal Society of Canada, made an honorary member of the Czech Entomological Society, and, in 1996, received the gold medal for research from the Entomological Society of Canada.*

Erik Tihelka: Where did your passion for paleontology get its start?

Jarmila Kukalová-Peck: I was born in 1930 in Prague, the beautiful medieval capital of the former Czechoslovakia (the present-day Czech Republic). My father owned a grocery store, and my mother had a dressmaker's shop. When I was fifteen, I saw a movie about a group of paleontologists excavating fossil dinosaurs in the Gobi desert of Mongolia. I was absolutely smitten and, when we left the theatre, I shouted, "I - want - to - dig - fossils." The image of people, waving hammers and working in a quarry, never vanished from my mind. I couldn't believe anybody could be so lucky as to do such interesting work and get paid for it.

In a small country like Czechoslovakia, universities could educate only a limited number of students. I originally applied to study English, but there were forty positions available for 800 applicants, and I was not accepted. I received this news only two hours before the registration deadline.

I hurriedly took a streetcar to Charles University and headed for the Faculty of Biology. It was located at the top of a long, massive staircase in a building that stood at the end of a steep street. I burst breathlessly into the building and accosted a tall man in a white lab coat. "It is very late, and I love biology. Could you please tell me where to sign up?" I pleaded. Though I didn't know it then, I was on the floor of the Faculty of Geology. The man looked at me and said, "Yesterday, we filled our quota of sixty new students. Run to the second floor, knock on Room 42, and tell the secretary that you want to study paleontology." That messenger was my future professor, Dr. Zdenek Spinar, a leading specialist in fossil amphibians and frogs, who opened up the gates of heaven to me. I went in with starry eyes and have lived happily ever since.

Studying paleontology was stunningly exciting. My new schoolmates (who were over-quota rejects from the faculties of chemistry, physics, mathematics, and other fields) turned out to be the most gifted and hardworking class the Faculty of Geology had ever hosted. Out of sixty graduates, only three were listed on the "red index," meaning students who had received straight As: Ivo Chlupáč (a future university professor), Vladimir Satran (a future Czech ambassador to Africa), and me (a future Fellow of the Royal Society of Canada).

ET: Your earliest publications focused on Paleozoic insects of Czechoslovakia, including the descriptions of many new species mostly belonging to now extinct insect orders and even the earliest true beetles.

JKP: As a budding paleontologist, I had to work with locally available fossils. I chose Paleozoic insects because some of their organs, wings, mouthparts, and genitalia were unresolved evolutionary mysteries. I loved Charles University and became an assistant professor. The head of our laboratory, Mr. Karel Havlata, and his wife Anna, helped me discover a splendid new locality of Paleozoic insects in Moravia, near the village of Obora. It was located in a 300-year-old road cut, used by the stagecoaches running between the Baltic Sea and Italy. I bought a discarded streetcar to sleep in, had it placed on the site, and the Havlatas, my students, and I used it for two months every year from 1956 to 1970. Obora eventually became one of the world's five richest localities of Paleozoic insects, and some of them are described in my papers.

Professor Kukalová-Peck and a group of her students from Charles University collecting Permian insect fossils at the north-western slope of the Hradisko hill in Moravia, Czech Republic, in 1954.

ET: Is there anyone who had a special impact on forming your research career?

JKP: When I was 33 years old, I received a letter that changed my life. It was from Dr. F. M. Carpenter, a professor at Harvard University and a specialist in fossil insects. He offered me an American grant to spend two months working in his laboratory. By that time, the political oppression in my country had diminished due to Alexander Dubček's "socialism with a human face" reform movement, and I was allowed to go.

Once in America, I adored everything: scientists, fossils, the generosity of the people, freedom of speech, you name it. At the end of my time in Carpenter's lab, I returned to Prague, and, about three months later, I received a peculiar letter from The Ministry of the Interior. I was to report to them in secret. When I entered the assigned room, it was entirely dark. Two men sat at a large table, and they aimed a bright light, a sort of standing lamp, directly in my face. They informed me that, while I was at Harvard, I had been followed by an agent who heard me criticizing the Communist government. If I confessed and sincerely apologized, I would be forgiven and could return to work. If not, they were aware that I was a single mother with two small children. If I did not cooperate, I would have only myself to blame for the consequences to their futures.

This went on, with some nuances, for the next four hours. I was in abject terror. Very obviously, they could do anything they wanted to me and my children. At Harvard we discussed insects, not politics, and when I was quite rarely asked how we lived under Communism, I gave truthful, not necessarily flattering answers. I had no suspicion I could be speaking to a Czech spy. What saved me were Agatha Christie's descriptions of police interrogation methods. I was clearly a political nobody, not worthy of the time and money it would have taken to send an agent after me, so I vehemently denied the accusations and claimed to be a devoted citizen.

Still, the interrogation continued. Perhaps the

trick was to get angry? The thought was so tempting that I jumped up, banged the table with both fists, and yelled, "Bring the agent in, I want him to look into my eyes and repeat his false accusations." Once I started, I couldn't stop myself, and I overturned the heavy table with a thud. My inquisitors immediately switched off the lamp, asked me to sign a statement saying that I had not been physically harmed, and in ten minutes I was on the street. I learned eventually that many other scientists experienced the same faked accusations after having visited Western countries. Some of them were punished and even imprisoned.

During the 1960s, Dr. Carpenter invited me several times to help him with the book on fossil insects that he was preparing (it has since become a basic textbook). I was making new interpretive drawings of the most important specimens from the Museum of Natural History in Paris and the Soviet Academy of Sciences in Moscow. Then, in 1968, just as Charles University was hosting the International Geological Congress, another political disaster struck my country: the Warsaw Pact invasion.

The sky filled with thundering warplanes, and in the morning tanks were everywhere in the streets. The USSR had invaded, hoping to quash Dubček's movement. The Geological Congress abruptly ended, foreign visitors fled in horror, and quite a few Czechs left along with them. By then I had an invitation to work for a year at Harvard and I became anxious about whether I would be allowed to go. Ernst Mayr, director of the Museum of Comparative Zoology at Harvard, promptly came to my rescue. He invited me to become an Alexander Agassiz lecturer in the Department of Biology, and I left Prague in the fall of 1968 for a year at Harvard.

ET: What were your strongest impressions from your time at Harvard?

JKP: At Harvard I had a room in the beautiful, luxurious Radcliff College, which offered near-gourmet meals and bathrooms where the towels were changed daily. My task was to research a large collection of 300-million-year-old fossils of gigantic insects from France. They belonged to the extinct order Palaeodictyoptera and are known for their long, beak-like mouthparts. I worked so hard that I skipped lunches and ate cold suppers late at night. Life was amazing.

In Carpenter's laboratory, I shared Thursday afternoon teas with E.O. Wilson, often met James Watson (of DNA fame) in the hall, and picnicked several times at Ernst Mayr's rustic cottage. Nevertheless, by June, my thoroughly overworked body started giving up. When Stewart Peck, who was finishing his PhD in beetles and biogeography with Professor Philip Jackson Darlington, invited me for an outing at the sea, I was overjoyed. We began seeing each other constantly, but I quickly became convinced our lives were going in completely separate directions and that it would never work. He was about to leave for Mexico, and I had two children waiting in Prague for their mama.

Stewart was upset, but after a day to think about it, he declared that he'd found a brilliant solution. I would wait a year for him to finish his doctorate, and then he would fly to Prague and marry me. The Czech government would then have to let us emigrate to America. The months that followed were a nightmarish mixture of responsibilities, dread about unexpected risks, and hopes for the future. I was emotionally so exhausted that I finally told Stewart marriage was definitely out. He merrily suggested we

*The Palaeodictyoptera were a diverse order of winged insects distantly related to mayflies and dragonflies that emerged in the Late Paleozoic. The wingspan of most species was 6-15 cm, but the largest known Carboniferous representative reached over half a meter. Their mouthparts were extended into unusual, stiff, elongated beaks presumably adapted for perforating primitive fruit and sucking out their contents. They became extinct together with the seed ferns, Cordaitales (an order of woody plants reminiscent of modern conifers), and other Paleozoic plants that probably were their main sources of food. Remarkably, the Palaeodictyoptera were the only major insect group that succumbed to the Permian mass extinction. The rest of the Carboniferous higher taxa are still with us today. Pictured: Reconstruction of a Carboniferous palaeodictyopteran (*Homaloneura *sp.) feeding on a *Cordaites *cone (from Shear & Kukalová-Peck, 1990).*

give it some time and, a few months later, we got married just before he left start a new teaching job at Carleton University in Ottawa.

When I went to the Canadian consulate to begin the emigration paperwork, the bureaucracy threatened to finish me off. I was asked for documents I couldn't possibly produce, and the stress became enormous. My last option was to try to get help from the US consulate, but I decided to use the "back door." I ambushed a sympathetic looking secretary and asked her for advice. She told me that we could get US visas immediately, but only if both Stewart and I swore to the authorities that, after a month, the kids and I would return to Prague. My face fell, but she winked and added. "Instead, you'll drive to the Canadian border, show your papers to the Americans, and ask for landed immigrant status. You'll have it about half an hour." ("Landed immigrant" was the term in use at the time; now, we just call them "permanent residents.")

When I told my American husband that he would have to swear to send his new family back to where it came from, he thought I had lost my mind. Then he saw the inevitable. When it was finally all over, we met in Boston—both certified criminals for the affidavits we'd signed and the oaths we'd taken, never intending to keep them. Obviously, even democratic governments can throw mind-boggling hurdles in front of people seeking citizenship.

ET: What happened to your family and research after arriving in Canada?

JKP: Stewart and I have very similar preferences, habits, and goals, and thus are well suited for each other. For forty-nine years we shared the same work room at home and at Carleton and never stopped enjoying it. As a biogeographer, he specializes in soil and cave faunas, and is known as an outstanding field biologist. For almost fifty years now, we've collected insects each summer and winter. During university sabbaticals we collected the entire year. I assisted Stewart as much as I could, and when the traps were busy catching insects, I sat in local biology stations and museums and studied insect wings.

We collected insects on all continents except Antarctica, but that was only be-

Above: *The oldest known giant mayfly,* Bojophlebia prokopi, *from the Carboniferous of the Czech Republic, described by Professor Kukalová-Peck. Scale bar = 4 cm. (Image courtesy of Pavel Sroka.)* Below: *One of the earliest beetles in the world, the iconic* Moravocoleus permianus, *of the extinct suborder Protocoleoptera. This specimen was collected at Obora, Czech Republic, by Professor Kukalová-Peck and Sarka Značkovská. (Image courtesy of Jakub Prokop.)*

American Fossil

*The Only Quarry that lets you keep every fossil you find! **

Full Day Dig (8 Hours) - $120
4 Hour Dig - $85
1 Hour Dig - $35
Children 7-14 Half Price with Paying Adult
Children under 7 Free with Paying Adult
Special pricing for groups over 5
Call or email for info 801-836-7269
Prices Subject to Change

Quarry Lease Opportunities Available

7 Days, 10 Days, 30 Days or Full Season

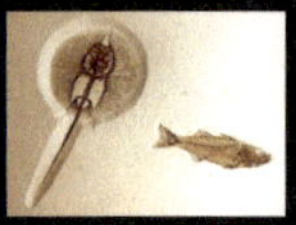

American Fossil is providing you the adventure of a lifetime. We have the only quarry in the world where you have the opportunity to dig fossils and keep every fossil you find. Our quarry is located in the world famous Green River Formation of Wyoming and is the most exciting destination in the world for everyone who loves fossils. Think of it as your own fossil expedition.

Ours is a world renowned quarry, well known for its fossil fish, and one of only two locations in the world where a complete 3-Toed Horse has been found. At the quarry you can also find stingrays, shrimp, Gar, Paddlefish, Amia, Pike, Dioplomystus, Knightia, and even turtles, crocodiles, bats, birds, ancient rhinos and camels, primates, and much more.

*At our quarry you can keep every fossil you find up to $100,000 in individual value. This means keep all of the turtles, bats, birds, stingrays and fish you find. No other quarry lets you do this. Come share our love of fossil collecting and contact us today!

www.AmericanFossil.com or www.FishDig.com

801-836-7269

cause I turned down an invitation to go there. I collected in about a hundred "wild" caves, and Stewart in many more. We used many kinds of flight-intercept traps and black lighting traps (attracting insects to lights at night), sprayed insecticide on logs, sifted forest litter, beat bushes over sheets, processed termite nests, and set many types of baited insect traps. Once we retired, we donated most of our modern and fossil insect collections to the Canadian Museum of Nature in Ottawa.

By doing insect research without teaching, meanwhile, I had much more time for fact-finding and problem-solving. I decided to go for something complex and exciting: the unresolved origin of insect wings and powered flight.

ET: And, in fact, you became an authority on the evolution of one of nature's most successful structures—the wings of insects. Why do you choose to examine the ways in which insect wings evolved and diversified?

JKP: In Canada, my research on insects took a special twist. Entomologists almost always work with insects at the species level, while the origin, structure, and evolution of the organs, mouthparts, wings, and genitalia receive much less attention. Interpretations are often approached with surprisingly little solid data and much speculation. Forty years ago, organ origin and structure were quite unresolved. I became interested especially in the wings. As a paleontologist, I was aware that they could be rich in evolutionary information, but only after all of veins and sclerites (the hardened parts of arthropod exoskeletons) in all insect orders had been derived from a single shared ancestral proto-wing organ. Only then would we be able to identify meaningful differences, similarities, and evolutionary relationships. When I came to Canada, wing veins in different insect orders were often named differently, wing sclerites were not even interpreted, and a phylogenetic tree of insects based on wings did not exist. This was a major hole in our understanding because pterygote (winged) insects originated and diversified with their wings, and powered flight was responsible for their survival and unparalleled success.

ET: When did flapping insect wings originate? Is it true that it happened with the appearance of tall trees, at about the middle of the Paleozoic Era?

JKP: Definitely not. Ancestral wing appendages were originally small, while forward flapping is very complex and only possible with large wings and well-coordinated leg muscles. Everything must have started very early. Insects diverged from a crustacean stock, and the oldest known fossil Crustacea, known from the beginning of the Paleozoic, were aquatic. Winged insects were first recorded much later in the Carboniferous, when they were already advanced and had diversified into most of the higher taxa found today. Because fossil insects with proto-wing organs remain unknown, the proto-wing from which all fossil and modern wings evolved had to be reconstructed. Unless that was done, the relationships among wing veins and sclerites from different modern groups could not be determined, and their characters could not be compared and evaluated to show phylogenetic relationships.

The iridescent wing of a living derbid planthopper. (Image courtesy of Melvyn Yeo.)

Insects are a huge group, and showing homologization (correspondence in structure and in origin) took me many years. I did it mostly during field work in the tropics where wing homology became a captivating part of my fieldwork. After almost fifty years of wrestling with comparisons I am quite confident that all major homologization problems are now solved. Homologized veins and especially sclerites offer a plethora of previously unavailable characters.

ET: What approach did you use to study the origin of insect wings?

JKP: The origin of insect wings has been a puzzle for more than two centuries. Gliding flight is easy and frequent, but Nature's gliders—frogs, lizards, snakes, opossums, squirrels, phalangers (marsupials with webbed toes), etc.—never developed powered flight. Flapping flight requires very complex movement, which can be generated only by thoroughly specialized leg muscles. The rule is that all "flappers" can glide, but no "gliders" can flap. Nature's flappers include extinct pterosaurs, birds, bats, and insects. All of these animals had to "sacrifice" their legs to acquire flapping wings except for insects. Insects kept their legs, and evolved two pairs of large, efficient thoracic flapping wings.

How to find out from which body part, and how, such "miraculous" thoracic wings originated? A paleontologist would start working from the earliest known ancestor.

How animals acquired the ability to fly is a question of major interest to evolutionary biologists. Flight enables an animal to disperse across physical barriers, live in comparatively safe places away from ground-dwelling predators, access food sources that others cannot reach, and find mates. Vertebrates evolved the ability for powered flight independently in several lineages, such as pterosaurs, birds, and bats. But insects got there first, and it is widely believed that they evolved wings and the ability to fly only once.

Debate on the subject goes back to the nineteenth century when there were two main positions on the origin of insect wings: whether they developed as entirely new structures or from pre-existing structures. As the latter idea gained acceptance, the discussion moved on to which structures had become wings. When *Lithomantis carbonarius*, one of the first articulated Carboniferous fossil insects, was discovered, support grew for the theory that wings were a development of the dorsal part of thoracic body segments.

Lithomantis belongs to the extinct order Palaeodictyoptera, widely regarded as the most primitive group of winged insects. Its oldest known fossil representative (the oldest known winged insect), is the 325-million-year-old *Delitzschala bitterfeldensis*. Palaeodictyoptera had their heyday in the Carboniferous, rapidly diversified into over thirty families, and were extinct by the end of the Permian (252 million years ago). They had a pointed beak for piercing and it is believed they fed on fluids from the seeds and stems of giant clubmosses and tree ferns. Palaeodictyoptera possessed two pairs of outstretched wings and veined lobes (which can be seen at the upper right of the specimen of *Lithomantis*, below).

In the first half of the twentieth century, there were two main theories for the origin of flight in insects, the "flying squirrel" and "flying fish" theories, depending on whether the author believed flight originated on land or from the surface of the sea. Both hypothesized that extended thoracic lobes enabled insects to glide, then to steer, and finally to flap. The "flying squirrel" model became generally accepted, particularly supported by the discovery of the then-oldest known fossil "insect," the 407-million-year-old springtail, *Rhyniella praecursor,* preserved in Devonian Rhynie Chert.

In the 1970s and 80s there was a renaissance of discussion regarding the origin of flight in winged insects. One important suggestion was that the greater surface area provided by those extended lobes could have been used for thermoregulation, i.e. the quicker an insect warmed up by basking, the more active it would become, and it was argued that this was a driver for the lobes to extend in the first place.

Nymphs of the Palaeodictyoptera (an ancestral group of primitive Palaeozoic insects with wings) supported the theory that the ancestral pterygote had three pairs of veined thoracic lobes, enabling it to warm up and become more active faster. Extended lobes with some articulation enabled the insect to glide. The final step was for the articulation and musculature to develop to allow two pairs of wings to flap actively. This major development allowed insects to colonize the world.

Excerpted from Andrew Ross's "Insect Evolution: The Origin of Wings" (6 February 2017). Current Biology *27(3): R113-R115. Available at: https://doi.org/10.1016/j.cub.2016.12.014.* Figure at left: Lithomantis carbonarius, *a palaeodictyopteran from the Upper Carboniferous of Scotland. Scale bar 10 mm. Photograph by Phil Crabb. Copyright Natural History Museum, London. Taken from Ross (2017).*

Insects and the related Crustacea are arthropods, and insect wings are powered by leg muscles. Hence, a proper Grand Evolution investigation should commence with the ancestral arthropods and their limbs.

ET: Which characters, in ancestral arthropods and their limbs, would be relevant to unravelling the origin of insect wings?

JKP: Ancestral arthropods are related to segmented worms. Thus, their heads, thoraxes, and abdomens were composed of many identical body segments, each bearing a pair of identical legs composed of the same number of limb segments. This means that arthropod legs are serially homologous—that is, insect mouthparts, thoracic legs, abdominal leglets, and genitalia evolved from the same kind of ancestral leg. The leg segments were originally cylindrical, but, in the upper leg, one to three segments sometimes flattened into a plate-like structure called a pleuron, which supported the sides of the body. The evidence suggests a strong possibility that insect wings evolved from just such a structure, flattened into a pleuron, in the first thoracic leg segment—called the epicoxa.

ET: What other evidence shows that insect wings originated from the first limb segment, and not from lobes or other structures in wingless insects?

JKP: There is important evidence. (1) the position of the wing buds, (2) the development and articulation of juvenile winglets, (3) the articulation of adult wings to the tergum (the "top" or dorsal portion of an arthropod segment), and (4) the location of the thoracic wing homologue in primitive wingless modern insects (Archaeognatha). Wing buds remain in some living juvenile dragonflies, beetles, flies, and caterpillars. They always occur far from the dorsal tergum in the "leg region," and then move upward, with each exoskeleton molt, until they reach the dorsal tergum and fuse with it. This clearly shows wings as independent from the tergum. Most Carboniferous nymphs have articulated and movable winglets that fossilized separately from their bodies—Permian rocks, for example, sometimes contain hundreds or even thousands of isolated mayfly winglets—but winglets are fused with the tergum in some fossil nymphs and absolutely all modern nymphs. Again, this pattern clearly shows that all these fusions are secondary, and that they evolved many times in parallel.

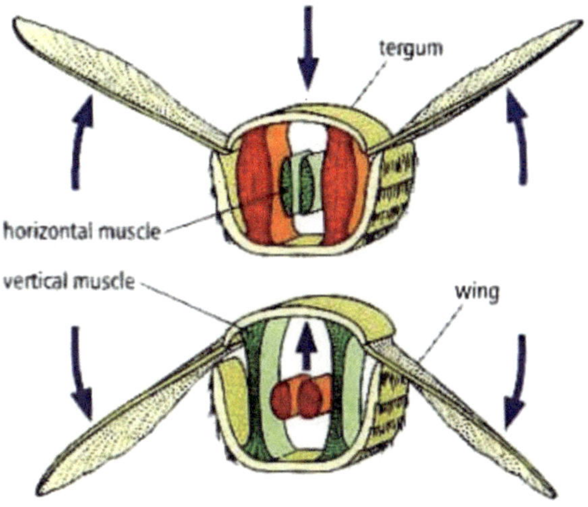

This very simplified diagram shows the tergum, the upper or dorsal "plate" of a thoracic segment, and its wing attachment, along with a representation of the complex muscles that allow the wings to move up and down.

ET: Based on the evidence you have examined, what is your final interpretation of the insect wing? When and how did the insect proto-wing evolve for powered flight, and what does it mean for insect phylogeny?

JKP: Wing sclerites originated from the epicoxal pleuron, and the basic adaptations necessary for aerial flight included: (1) the strengthening of the anterior wing margin by the fusion of veins, (2) reinforcement of the long wing axis by several areas of veins associated or fused with one another, especially near the wing base, (3) reinforcement of the wing base braces between veins and/or hardening of skeletal elements, and (4) the existence of groups of muscled "super-sclerites" that improve wing flapping and flexing. Importantly, the flight-enabling characters in higher insect taxa have remained almost unchanged for 300 million years, are present in all members, and provide the evolutionary-morphological data necessary for reconstructing the insect phylogenetic tree.

Some crustaceans and the wingless insects, when disturbed, start to hop. In the Lower Paleozoic, there were no trees to climb. Insects had many predators, and, to escape from them, they were probably jumping rather than merely walking. Mobile, flexible extensions of plates on the upper part of the thorax probably helped them improve hops. With time, this proto-wing enlarged and muscles became better coordinated, and hopping was abandoned for aerial flight. The emergence of tall trees in the mid-Paleozoic (Devonian) very probably accelerated this development. The oldest winged insects, which are known from the Carboniferous, were already very advanced and diversified into all modern superorders. Winged insects very probably started to diversify quite early, possibly during the flight-hopping stage, long before aerial powered flight was adapted. In fact, insect higher taxa in general probably started diversifying before aerial flight was established.

ET: Your system of insect wing venation has become widely used among entomologists and paleoentomologists working on different groups. Hypotheses regarding the origin of insect wings have always attracted controversy, though. How has this affected your work?

JKP: Scientific interpretation of insect organs must start with their earliest known ancestors, which are the primitive arthropods. Extremely broad comparative methods involving distant animal groups and different kinds of organs, which are used by evolutionary morphologists, or piecing together evidence from widely scattered sources are approaches that some entomologists find quite suspicious. They instead study the diminutive and often variable body details under the rule of "what I see is there" and evaluate them mostly cladisticly. From this angle, reconstructing long-reduced features with the help of inscrutable fossils seems unreliable.

The first researcher to give a sound scientific basis to organ interpretations was Alexander Grigorievich Sharov (from the Russian Academy of Sciences in Moscow). In 1966, he published the book *Basic Arthropodan Stock, with Special Reference to Insects* for which he was extremely and unjustly criticized by many of his colleagues. After 1970, Sharov's research was superbly developed and advanced by the extraordinary morphologist Edward Laidlaw Smith at the California Academy of Sciences in San Francisco. Over twenty-five years of daily work, Smith gradually resolved all morphological structures of the insect head, thorax, and abdomen, except wings. He wrote a huge, coherent, informative, and expertly illustrated manuscript offering copious, detailed, and easily testable evidence. His manuscript was criticized so strongly by official reviewers that he withdrew it from publication. I collaborated with Smith for years, sharing information with him on Paleozoic mouth parts and genitalia, and I used and verified his evidence. I applied and cited many of his interpretations in my papers. They are now the only record of his monumental work. On the brighter side, some of his theories are now broadly accepted. The manuscript of his book is in my files and, after I am gone, it will go to a museum interested in advancing Grand Evolution research. Who knows, Smith's book may eventually even be re-edited and published.

Science is extremely competitive and some scientists, mostly male, are personally insulted by differing opinions, especially if they come from a woman who doesn't use cladistic methods. But I made mistakes, too, and criticism is necessary. Still, there are limits. Women in science are often criticized more harshly than men, and sometimes with patronizing overtones. In my case, critiques were often not scientific but personal and very impolite.

Anyway, as we used to say in Prague, "One doesn't drift through life like a feather." During my fifty years of comparative morphological research, I made at least 20,000 documentary wing figures covering all insect orders. My intention was to use them to construct a phylogenetic tree of the higher taxa of insects. Midway through this task, though, I realized that finishing it properly might take another thirty years. Luckily, humanity invented an efficient dispenser of big scientific data—the internet. I've selected a group of 800 of my most informative insect-wing figures from all orders and color-coded their homologous veins and sclerites. They will be donated to a museum that will make them available online to entomologists interested in using them in their own research projects, no strings attached. Except for one little exception, egotistic but perhaps understandable: I want to be cited as the interpreter and homologizer of pterygote wings and sclerites based on the scientific evidence.

With a project like that, I have no time to waste on disputes. In science, truth always has an irresistible tendency to prevail, tomorrow if not today.

USPS Issues *Tyrannosaurus rex* Forever Stamps

With a pane of sixteen stamps that features four designs, the U.S. Postal Service brings *Tyrannosaurus rex* to life—some 66 million years after its demise.

Tyrannosaurus rex dominated the tail end of the dinosaur age. A recent surge in discoveries has revolutionized understanding of this mighty carnivore, often described as the "top predator" of its time.

Four dynamic designs on a pane of sixteen stamps depict *T. rex* in growth stages from infancy to adulthood. One design illustrates a *T. rex* in a forest clearing; another shows the same adult *T. rex* with a young *Triceratops*.

The third and fourth stamps depict a newly hatched *T. rex* covered with downy feathers and a bare-skinned juvenile *T. rex* chasing a primitive mammal.

"The Nation's *T. rex*," the young adult depicted on two of the stamps, was discovered on federal land in Montana and is one of the most studied and important specimens ever found. Its remains are now exhibited at the Smithsonian Institution's National Museum of Natural History in Washington, DC. Art director Greg Breeding designed the stamps with original artwork by scientist and renowned paleoartist Julius T. Csotonyi.

Two of the four stamps feature a lenticular overlay, causing the bashful *T. rex* in a forest setting to appear to lunge and roar, and the tender moment between young *T. rex* and Triceratops is shown as both life reconstruction and in skeletal, fossil form.

Nicklas Smyser and the first stamp canceled using his award-winning artwork. © Loudon Now.

As part of its rollout of the new series of stamps, the U.S. Postal Service sponsored a contest for artists to design a special, first-day-issue cancellation stamp.

The winner was Leesburg, Virginia, elementary school student, Nicklas Smyser, 8, described as "a qualified dinosaur enthusiast who can easily name dozens of them." Smyser was invited to join the postmaster of the Leesburg Post Office at a ceremony and to make the first cancellation with the new stamp based on his drawing.

The Tyrannosaurus rex stamps are "Forever Stamps," meaning they will always be equal in value to the First-Class Mail rate. They can be purchased online at usps.com/shop, by calling 800-STAMP24, by mail through USA Philatelic, or at Post Office locations nationwide.

The Mostly True Tales of a Commercial Paleontologist

HEAD FULL OF ROCKS*

An excerpt

from the forthcoming book of short stories by

Seth Sorensen

WANDERING

Not everyone who wanders is lost. Some of us are just searching for fossils.

I had spent the morning roving the hills and valleys of southern Wyoming in search of the remains of a fossil deposit that had been dug in the 1800s. The wind had dried my lips long ago, and the sun was slowly turning my skin to leather.

My friend David, who looked like a malnourished, brown-tipped cotton swab, and his single-toothed friend, Bert, had accompanied me. We were quite the sight, wandering through the brush and rocks, hunched over, staring at the ground, looking much like someone scouring the area for loose change to pay his bills.

"Hey," David shouted over the hill at me. "Have you got a couple dollars?"

"What do you need a dollar for?" I hollered back.

"I'm thirsty and want to buy a soda," he shouted back.

"You do realize that we're out in the middle of nowhere and there isn't a place you can buy a soda for fifty miles, don't you?" I bellowed back.

"What's your point?" he asked.

"My point is that you are a bit loopy," I said back.

"Hey," he shouted again, "that's all part of my charm. If you don't like it, then I'll stop, and you can get off my unicorn. Have you got a couple of bucks or not?"

I just rolled my eyes and continued walking.

"You know what?" he went on.

"Let me guess," I responded, "You're going to stop talking and let me walk in peace?"

He went right on as if I hadn't said a word, "Remember when we were kids? We could go to the store with a dollar and come home with two bags of chips, three or four candy bars, some sunflower seeds, and a cold drink. It was awesome. Nowadays they got cameras everywhere."

"Please stop talking," I sighed in response.

By late afternoon, we were just about to give up when I found something that looked like fossilized bone. Doing what any great paleontologist would do, I picked it up and immediately, without cleaning it off, placed it on my tongue. It stuck.

"I found bone," I shouted excitedly.

The other two came running over immediately and together we began following a trail of small bone shards up a wash and onto the side of a large hill. Suddenly there it was, the old quarry.

The old dig site was now overgrown with sagebrush and weeds, but there was no mistaking it. It was a large depression in the ground, rocks and rubble strewn about the nearby hill. There were also random bits of rusted metal and a few scraps of what appeared to be burlap.

We were elated. This quarry had not been seen since it was vacated over a century ago. We set to work carefully photographing every minute detail, right down to the smallest fragment of stone. We were careful not to disturb anything because the site was a treasure and needed to be preserved for future generations.

After documenting the entire site carefully through photographs and detailed descriptions, we set off back to the truck. We were certain the discovery would be big news in the fossil community, and were all thrilled.

On the way back, David looked at Bert and asked, "Have you ever done any online dating?"

"Nope," replied Bert, "I would rather see the person. It is pretty hard to have dinner or go to a movie if you have to bring your computer."

"Don't be crazy," replied David indignantly, "That just means you find someone to date online, and then you meet them in person, not just date them over the computer."

"Oh," said Bert, looking at his feet. He was not a great conversationalist.

"I've been working on an online dating profile," David went on. "I'm thinking something like: Grew up in the Jurassic; enjoy living a sedimentary lifestyle; all original teeth, no restoration. What do you think?"

"I like it," said Bert, looking at David out of the corner of his eye.

"You do know that's only funny to someone that understands fossils, right?" I asked.

"You don't know anything," was his reply. "By the way, you're welcome for the get better soon card I left in your trailer."

"You wanted a thank you for that?" I asked. "I wasn't even sick!"

"I didn't give it to you because you were sick," he said. "I got it for you because I just think you can do better."

Sometimes our conversations seem to make absolutely no sense at all.

LOST

Let's start out with a little cowboy poetry:

> *Roses are red, violets are blue,*
> *if you find a cow pie don't eat it,*
> *it's poo.*

I realize this is neither true cowboy poetry nor even poetry at all, but is a necessary element to get this story kicked off.

Growing up in the west, my childhood had been filled with stories of the adventures of early pioneers and their ingenuity. They traveled the plains in the dead of winter, the temperatures well below zero, in a quest to find a better life. As they sat freezing on the frozen ground with no wood to burn, they realized that "Buffalo chips" (dried bison dung) could be used as fuel for their fires. The droppings were plentiful because of the massive herds of these giant animals roaming the prairies.

I had always imagined myself vagabonding across the wide-open spaces, a pistol strapped to my hip and a rifle sling over my shoulder. My horse was a giant, its hooves leaving craters in the ground as it ran. We would stop at night and have a dinner of fresh bison leg, cooked over a roaring fire of buffalo chips.

In real life, David and I would have to relegate ourselves to hot dogs cooked in a coffee can that perched atop a stinking, smoldering pile of cow pies that we'd harvested from the pasture behind my house. The smell was pretty bad, but the flavor it imparted to the hot dogs was worse, although neither David or I would admit it.

Much older now, we often sit in the lip of the quarry, cooking hot dogs over a fire built from sagebrush and old palettes. As it happened, we were doing this one night when David jumped to his feet. His hot dog dropped into the fire, and his chair shot backward.

"What is that?" he exclaimed, pointing off into the distance.

I gazed in the direction of his finger but didn't see anything.

"What are you talking about?" I asked.

"Right there!" he said, agitation creeping into his voice. "That light over there on that hill!"

I looked again, my eyes adjusting to the dim light, and I noticed a very faint light bobbing up and down several miles in the distance.

"I see it now," I replied, squinting.

"I think it might be a UFO," David whispered, panic now entering his voice. "It might be a ghost, too," he continued, "or a ghost alien!"

"Give me a break," I said, rolling my eyes.

"Quick," he whispered again, "put the fire out."

"I am not putting the fire out," I replied.

"We need to put it out!" he said, now frantic. "The ghost lights will see us!"

With that, he began throwing fistfuls of dirt onto the fire. The flames gradually expired, leaving us in darkness, the only light coming from the stars overhead.

As we continued to watch the light, we realized that it was moving closer. It was now probably just over a mile away.

The light began to flicker and then went out. "Where did it go?" David asked. "I've heard that ghost lights go out right before they attack."

"Don't be stupid," I said. "That's nothing mysterious. It was probably just a firefly or something." I was feeling a bit uneasy at this point.

"Shh," David said quietly, clamping his hand over my mouth. "Do you hear that?"

I strained my ears and, as it turned out, I could hear a faint cracking sound in the distance.

"Ooohhh," whispered David, "I wish I had my stainless-steel salad fork right now. I never go anywhere without it, and the one time I don't have it, this happens."

"What are you talking about?" I asked, very confused.

"Stainless-steel salad forks are the only known thing that keeps ghost lights away," he replied in a tone that indicated he thought I was a total buffoon for not knowing that.

The crashing sounds grew louder and closer. It sounded like it was at the base of the hill leading up to the quarry. Those sounds were now accompanied by a loud moaning sound and the occasional growl. I began to think David was right. Maybe this was a ghost light about to attack.

> Stainless-steel salad forks are the only known thing that keeps ghost lights away....

On hands and knees, we crept quietly away from the fire pit, keeping our eyes pinned to the lip of the hill to see if anything crested and became visible in the dim starlight.

"Ahh," screamed David, "It's got me!"

This was followed by a loud crash and the sound of a landslide of metal and glass hitting the ground. He had backed into our picnic table, spilling everything.

Fearing the ghost light had heard us, I flattened myself in the dirt and tried not to breathe. Another moan floated through the air, and a figure appeared in front of us. It was not a ghost light, but a zombie. It limped slowly toward us, moaning and hissing.

David jumped to his feet and threw a large rock, hitting the creature in the head. It dropped to its knees, holding its head in its hands, and whimpered.

"You can't have me!" David screamed and raced forward, somehow having found a large board. He brandished his new weapon like a ninja, swinging it back and forth in an attempt to chase back the zombie.

"Stop! Please, stop!" begged the creature in a surprisingly human voice.

I pulled my phone from its case and lit the flashlight. There, hunkered down in the dirt, was a man caked in dry mud, his clothes in tatters, cuts covering his arms and legs. He had a small white dog tucked under his arm.

"That is not a zombie," I told David.

He stopped swinging the board and started at the poor

man, now bleeding from a gash in his head.

After we'd helped the man to his feet and cleaned him up a bit, he informed us that he'd been cruising the hills in his jeep when he'd become stuck in a large mud hole. He'd spent most of the day trying to get it out with no luck. As it grew dark, he knew he needed to find help and set out walking with an old flashlight. He spotted our campfire and headed in our direction to find help. A little way into his journey, the light had gone out, leaving him wandering in the dark. He had fallen into small holes, tripped over bushes, and twisted both ankles on unstable patches of rocks. After walking for miles, he was a mess.

We loaded him into our truck and headed out to find his jeep. We located it just before 4:00 am and were able to pull him out quickly. He thanked us and headed off. As we got back into my truck I happened to glance out of my window and spotted another light bobbing up and down in the distance.

"What are you looking at?" David asked.

"Nothing," was my response. I put the truck in drive and headed in the opposite direction.

SPEED LIMIT

Spinosaurus is my favorite dinosaur. It was huge, had some of the longest claws of any dinosaur, and sported a bizarre sail on top of its body. The sad thing about having a favorite dinosaur at my age is that nobody cares, and they also don't care that I can pronounce hundreds of different dinosaur names. They just don't care.

Excitement was running high at the quarry. We had guests from one of the largest museums in Asia. We knew they were arriving with a passion for fossils and large bank accounts. We wanted this to be a special experience for them.

Because I wanted everything to be just right, I stayed at the quarry and sent David to collect our guests. Even though he didn't have a lot of experience driving on the freeway, I thought he could handle it.

David headed down the road to pick up our guests in the next town over. This was a journey of nearly 100 miles each way and took several hours. Nonetheless, David arrived at their hotel much earlier than anticipated, and he went inside to track them down.

Mr. Ling and his partner were sitting in the lobby, chatting. Recognizing them from their pictures, David approached them.

"You Mr. Ling?" he asked.

He replied that he was, and the two men accompanied David out to the truck. David opened the back door for Mr. Ling and gave him a slight bow as the man climbed in. Mr. Ling just gave him a quizzical look.

As they pulled out, David tried to start up a conversation. "I can tell you gentlemen how to get a smokin' hot body. Just walk around at our quarry in the middle of summer in a black shirt."

The two men stared at each other.

David tried again, chuckling to himself. "I can see you didn't like that one. How about this one. My wife thinks I'm crazy, but I'm not the one who married me."

The two men stared at each other again.

"That's funny because I don't have a wife!" David said between chuckles.

"Okay, okay. How about this one. Which is heavier, 200 pounds of feathers or 200 pounds of bricks?"

Mr. Ling just shrugged.

"Well," explained David, "most people will tell you they're the same but they are not. Two hundred pounds of bricks is just a bunch of bricks, but if you have to carry 200 pounds of feathers, you also have to carry around the weight of what you did to those poor birds, so feathers are heavier."

The two men sat in silence for the next hour of the drive as David continued to pelt them with jokes.

David pulled onto the interstate and, after only about a quarter-mile, he passed a parked police vehicle. That always made him nervous, and he quickly checked his speed. No problem, he thought to himself.

David suddenly saw the flashing lights in his mirror, as the police vehicle entered the freeway. David quickly pulled off to the side of the road, not sure what he had done wrong. He rolled down his window and placed both hands on the steering wheel in plain sight.

"Hello, officer," he said. "How can I help you? I can assure your officerness that I was not speeding, and that I was not using my phone or any other technologies that might impair my ability to properly navigate the automobile that I'm the legal and lawful owner of."

"Sir," the officer said, "you were not speeding, but you do need to be aware that driving significantly slower than the speed limit can be very dangerous not only to your safety, but to the safety of other drivers on the highway."

"Officer," David replied, still staring straight ahead, "I can assure you, and my passengers can testify, that I was going exactly the speed limit. I was traveling at exactly fifteen miles per hour."

The officer's mouth broke into a wide grin.

"Sir," said the officer. "You do realize that the signs with fifteen on them indicate that this is Interstate 15. The speed limit is seventy-five miles per hour."

A sudden expression of understanding dawned on David's face.

"Sir," the officer continued, "is everything okay? Your passengers seem shaken up and have not made a sound the entire time we've been talking. Can I do anything for you?"

"No," replied David. "Everything is great. They'll be just fine in a few minutes officer. The thing is, we just got off Highway 125."

As it turns out, we didn't sell a single fossil on that occasion.

AUNT BEV

When I was a kid, I used to love Sunday School. I learned many of life's most important lessons sitting there with my neighbors.

My favorite Sunday School teacher was Brother Jenkins. He was a seventy-year-old man with a long white beard, and he always wore red. To this day I believe he's the real Santa Claus.

One Sunday, Sister Jenkins was speaking to the congregation. She talked about a wide variety of rights and wrongs, followed by a heartfelt segment on family and relationships. She proceeded to talk about a time when she had been particularly tested.

"A few months back, Brother Jenkins and I went through one of the most trying times of our lives when he seriously injured his scrodum," she said to murmuring and whispers.

I had a hard time refraining from laughter, as did many others in the congregation, based on the large number of chuckles that quickly turned into coughs.

She continued, "Because of this injury we were not able to share the intimate moments that had made our marriage special up to this point. Even the hugs that we used to enjoy were extremely painful. We both missed the closeness we had shared for years."

She continued on for another ten minutes and then closed her talk. As soon as she began walking away from the pulpit, Brother Jenkins got to his feet in the congregation and approached the pulpit.

We were all amazed he would dare to just walk up after the story we'd all just heard. In a very composed manner, Brother Jenkins calmly pronounced just one word into the microphone. "Sternum!" he said, and he returned to his seat.

A few Sundays passed, and Brother Jenkins taught us a lesson on stories with a moral. I was very thoughtful as he shared a few with us. He asked us if we had any stories of our own we would like to share.

April, a prim-and-proper, teacher's-pet kind of girl with long black braids, raised her hand right away. "I do," she said, "My father is a farmer, and we have a really big farm. We have lots of laying chickens at our house. One time we were taking eggs to the farmers' market in an old basket. We put it on the front seat of the car to keep them safe. We hit that big pothole by the gas station and all of the eggs went flying. They all ended up breaking. It was a real mess that we had to clean up, and the car smelled bad for a week."

"That's a great story, April," said Brother Jenkins kindly. "What is the moral of the story?"

"Well," said April, "I think it's that you shouldn't put all your eggs in one basket!"

"Great," said Brother Jenkins.

Not to be outdone, Joey raised his hand.

"Go ahead, Joey," said Brother Jenkins.

"My family is farmers, too, and we have a big farm too," said Joey a bit mockingly, "but we don't sell the eggs. We raise chickens for the meat. We got sixteen eggs this year and kept them under the heat lamps. When they hatched we only got thirteen chicks."

"So what is the moral of your story, Joey?" asked Brother Jenkins.

"Don't count your chickens before they're hatched," he replied.

It was at this point that Brother Jenkins looked at me. I always hated it when teachers called on you because your hand wasn't raised. My brain raced through countless stories and dirty jokes that my father and grandfather had told me over the years. None of those seemed appropriate for church though. Then I remembered the story of my Aunt Bev.

"My dad told me this story one time about my Aunt Bev," I said as more of a question than a statement. "She used to live here in town."

"I remember her," said Brother Jenkins.

"Well, Aunt Bev flew a jet in the Gulf War, and one time her plane got shot down. She had to parachute out over enemy territory and all she had was a bottle of rum, a machine gun, and a machete."

Brother Jenkins raised his eyebrows and gave me a funny look, but he let me continue.

"Well, when she was falling, she drank the whole bottle of rum so it didn't break when she landed. She landed in a camp with 100 bad guys surrounding her. She killed seventy of 'em with her machine gun but then she ran out of bullets. Then she took out the machete and killed twenty more but that last guy broke her machete because he had a hard head. With no machete, she had to kill the last ten with her bare hands."

Good heavens," exclaimed Brother Jenkins, "what kind of moral did your father tell you from that horrible story?"

"Stay away from Bev if she's been drinkin'."

YOU CAN TAKE IT WITH YOU

About six years back, an elderly friend of mine came down with an illness. Ed and I had been friends for years, digging fossils and causing trouble all over the west. One day he called me and asked if I would mind driving out to Colorado to see him.

I loaded up my truck, and the next day I was on the road, flying cross country with the windows open and bugs in my teeth. I arrived at Ed's house shortly after dark. I walked up and knocked on the door, and his brother, Tom, answered, greeting me with a bear hug.

Ed and Tom were partners in their family's lucrative

fossil business. They owned a number of properties from which they excavated some of the largest dinosaurs on the continent.

"Ed's not doin' so well," Tom said sadly. "I don't think he'll be around much longer.

He escorted me back to Ed's bedroom where I found him lying in a cast-iron bed with a moth-eaten patchwork quilt made from old grease-stained pants. The quilt looked as if it had been worn out and stitched back together numerous times.

Ed motioned me to his side.

"I'm glad you're here," he said, clearing his throat and coughing into his hand. "I have something you need to hear."

This grabbed my attention. Was he about to reveal the location of a family of *T. rex,* or a dinosaur lodged in the door of a fossilized UFO (that had always been a dream of mine).

"Now, listen," he said in his gruff voice, looking at Tom. "And Seth you make sure he does this. When I pass on, I want you to take all of my money and put it in the casket with me so no one else can get their greedy hands on it!"

My jaw just about hit the floor. It wasn't some important bit of information he was going to share with me as his legacy. He simply wanted me as a witness to some fool idea.

"It should be about $180,000, and it's hidden in this blanket," he continued.

I looked at Tom, expecting to see shock on his face. To my surprise, Tom calmly agreed, and I could see the relief that brought to Ed's face.

"Now, you heard that," Ed said, looking me straight in the eyes.

The next day, much to our regret, Ed passed away. It was a solemn and sobering event for all of us. Preparations were made straightaway for the funeral. A beautifully carved wooden casket was selected, and arrangements were made.

The funeral was open casket, per Tom's direction. Ed was dressed in his favorite pair of tan coveralls, and his filthy, sweat-stained straw hat with the red bandana sat next to his head.

The service was beautiful. Ed, who was always one to try and avoid attention, would have hated every minute of it. Stories were shared of his life, his discoveries, and some of his crazy adventures. As the service was ending, Tom came forward to pay his final respects. Tom walked slowly to the casket, an old brown shoebox in his hand. I knew what was inside and still could not believe it. He lay the shoebox on Ed's chest, and the casket was closed and lowered into the ground.

I put my arm around Tom's broad shoulders and together we returned to his old pickup, neither of us saying a word.

When I couldn't stand it any longer, I whispered, somewhat more loudly than I had intended, "Were you really crazy enough to put all of that money in the casket with Ed?"

He looked at me and replied, "Of course, I did. I made a promise, and I always keep my promises. He was my brother, after all."

"So, you put every cent of his money in that casket with him? You're a much better man than I am."

"What can I say," he responded, a slight grin forming on his lips. I took all $180,000 and deposited it in my bank account. And then I wrote him out a check and put it in that shoe box."

Made in the USA
Las Vegas, NV
10 November 2026

57851948R00031